White Paper

Safety Solutions and Differences in Motor Vehicle Drivers Who Use Cellular Telephones

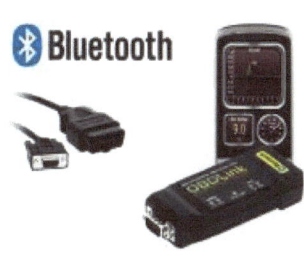
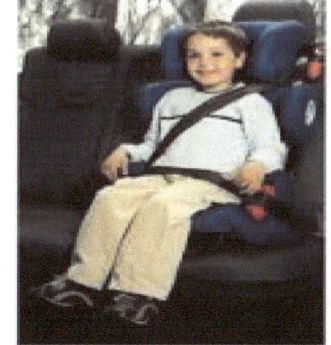

CONTENTS

Introduction

The Harvard Center of Risk Analysis reports "about 636,000 auto accidents occur each year because a driver was talking on a phone or text messaging" (Cohen, Ropeik, & Myron, 2000, p. 1). Cohen et al. (2000) estimated "2,600 people will die and 33,000 people will be injured annually" if legislation to regulate the behavior is not in place within each state (p. 1). A representative from the Illinois Department of Transportation said cell phone use while driving was the main cause of "at least 1,357 motor vehicle accidents in 2007 and resulted in 351 injuries and eight deaths" (D. Kline, personal communication, August 3, 2009). The numbers suggest driver attitudes and behavior must be monitored carefully to reduce world-wide deaths and injuries among motor vehicle drivers who use cell phones. Cohen et al. (2000) suggested driver distraction is a major public concern, and Cohen et al. calculated a driver's average risk of death while using a cell phone as 6.4 people per million per year. The "risk of a passenger, another motorist, or a pedestrian's death by a driver using a cell phone is 1.5 people per million per year" (Cohen et al., 2000, p. 2). This study explained the risks and dangers of cell phone use and the moderator variables used to determine how law enforcement efforts may play a significant factor in saving lives. As researchers continue to mine automobile accidents and cell phone use data, results provided further insight into human behavior. Given the stark statistics, this problem is a major concern for individuals, the government, and society, as well as telecommunications companies and their partnerships.

The general problem to consider is the growing concern for reducing the number of worldwide motor vehicle deaths (i.e. drivers use cellular telephones while driving); while the specific problem is in the raising of awareness and monitoring driver behavior and attitudes through state laws and police enforcement. Although the banning of voice and data (talking and texting) is common practice in some states, efforts to minimize driver risk are being considered through stringent laws nationally. The use of handheld devices while driving is a problem both for the driver who chooses to engage in behavior and for those who share the road with the driver. The increased and widespread use of cellular phones, PDAs (personal digital assistants), and other handheld devices to make calls and send text messages or e-mails (electronic mail) while driving has placed the public at new risks which are like those posed by driving while intoxicated. Though the number of related incidents has declined due to current legislation, it has not been enough to create lasting behavioral change and risk remains high (Mortazavi, Atefi, & Kholghi, 2011). For some individuals, traffic citations (both primary and secondary offenses) do not present a sufficiently strong deterrent, while others indicate when a police officer is not in the immediate vicinity, he or she will use their cell phones because he or she will not be seen breaking the law (Wang, Sato, Rau, Fujimura, Gao, & Asano, 2009).

One method of increasing driver safety is to educate drivers to influence attitudes, and thus, behavior using this quantitative study shown through a correlational research design shaped by empirical data with probability sampling. While some educational tools are available, there is a clear gap in the research on the relationships between attitudes about using cell phones while driving and actual behavior (Constant, Salmi, Lafont, Chiron, & Lagarde, 2009). Until this gap can be bridged, it is likely the morbidity [and mortality] associated with the use of handheld devices while driving will remain high.

The general population of the study addressed the growing concern of motor vehicle deaths within the United States (Northeast, South, Midwest, and West typologies; see Appendix B in main study). Although the study focused on the northeast area (population of a Boston area mall), the deaths in other parts of the country (i.e., Los Angeles) will need further research in the area and therefore be determined in future research. Therefore, it must be noted there is a bias due to the specific results of this study within the area of research. The total number of "deaths recorded between 1999 and 2005 suggests the concern for adding legislation will reduce the number of fatalities and injuries statewide and across the globe; the total number of fatalities is 311,356 over the same period" (Center for Disease Control and Prevention [CDC], 2010, p. 1330). To make a significant reduction in the national vital statistics, education and awareness of the problem must be communicated among driving populations (ages 16 and older).

The purpose of this quantitative research study was to investigate if there was a statistically significant relationship between attitudes about using a handheld cellular device while driving (independent variable) and actual use (dependent variable). This study also investigated if the pattern of relationship significantly differs by gender, age, marital status, parental status, ethnicity, age of children, level of education, or having a disability (moderator variables).

The study method and design chosen was quantitative due to the nature of the study variables (independent/dependent/moderator). This quantitative study explored and explained the dangers of motor vehicle drivers using cell phones with respect to gender, age, marital status, parental status, ethnicity, ages of children, level of education, and having a disability. The quantitative method took into consideration the amount, number, bulk, mass, measurement, and volume (capacity); purposefully engages in the critical diversity of deaths by motor vehicle drivers.

Understanding the ways in which attitudes influence behavior provided insight into the ways in which both education and legislation might influence drivers to limit or eliminate their use of handheld devices while driving. In addition, the specific examination of both gender and ethnicity as moderators provided additional insight into how educational messages might be refined to appeal to specific audiences. The pilot study was conducted around Boston, Massachusetts at an area mall and electronically while the full study was conducted electronically through a Survey Monkey audience. Respondent's demographic characteristics consisted of parents (both males and females) between the ages of 18 and older who use cell phones regularly. No other selection restrictions were instituted. The sample size determined appropriate for the pilot study was 30 while the sample size determined appropriate for the full study was 128 (158 individuals in total). The sample size was justified by the power analysis for the type of analysis used (see page 78). Further, sample size was also justified by the central limit theorem, which states the conditions under which the mean of a sufficiently large number of independent random variables, each with finite mean and variance, will be approximately normally distributed when sample size reaches 30 participants per group (Barany & Vu, 2007;

Rice & Trafimow, 2010). This implies a sample size of 60 participants total (30 per group) would be the minimum required.

Challenges in Safety Solutions

This study was guided by eight research questions, each of which are accompanied by their null and alternative hypotheses:

RQ1: What is the relationship between attitudes about the dangers of using handheld cellular devices while driving and reported dangerous use while driving?

RQ2: How does gender moderate the relationship between attitudes and reported dangerous use while driving?

RQ3: How does ethnicity moderate the relationship between attitudes and reported dangerous use while driving?

RQ4: How does parental status moderate the relationship between attitudes and reported dangerous use while driving?

RQ5: How does marital status moderate the relationship between attitudes and reported dangerous use while driving?

RQ6: How does the age of children moderate the relationship between attitudes and reported dangerous use while driving?

RQ7: How does having a disability moderate the relationship between attitudes and reported dangerous use while driving?

RQ8: How does having a higher level of education significantly moderate the relationship between attitudes and reported dangerous use while driving?

A quantitative method and correlational design were determined in both the full and pilot study. The basis for the pilot study was to test and obtain validity while the full study was to address the research questions. Data collection and analysis were appropriate choices to fulfill the purpose of this study, as the researcher was able to collect data from several individuals to analyze the relationship between independent/dependent/and moderator variables. This study was designed to identify interventions used by core business to positively affect levels of moral reasoning. The correlational study conducted focused on determining if the utilization of different approaches could positively affect levels of moral reasoning (Walsh, White, & Young, 2009). Onwuegbuzie and Johnson (2006) posited "a quantitative approach is appropriate for research focused on statistically testing hypotheses" (p. 60).

Data was collected from the sample demographics and survey instrument used to collect attitudes and perceptions. The data collection instrument was used in the pilot study to test validity and afterward in the full study instrument. The specific population of interest was Northeast drivers who are between the ages of 18 and older. A convenience sample was used, collecting data from individuals visiting a Boston, MA, area mall and posted to their email address electronically in both the pilot and full study [electronically only through Survey Monkey audience] instrument. The power analysis detailed in Chapter III indicated 30 participants were necessary to answer pilot study research question sections, while 128 participants were necessary to answer full study research question sections; thus, 158 total participants was required to conduct this research.

The Dangers of Searching for Information, Texting, E-Mailing, Reading, Downloading and Browsing

DRIVER ABILITY TO MANAGE THE AUTOMOBILE

Purple	-	Moderately Severe Incapacitation
Light Blue	-	Severe Incapacitation
Blue	-	No Incapacitation
Green	-	Moderate Incapacitation
Crimson	-	Minor Incapacitation

HOW MUCH EFFECT DOES USING A MOBILE COMMUNICATIONS DEVICE WHILE DRIVING HAVE ON DRIVING ABILITY – SPEAKING ON THE CELLULAR TELEPHONE DEVICE

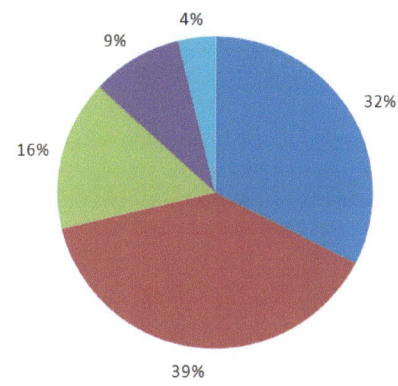

Searching for Information

HOW MUCH EFFECT DOES USING A MOBILE COMMUNICATIONS DEVICE WHILE DRIVING HAVE ON DRIVING ABILITY – SEARCHING FOR INFORMATION

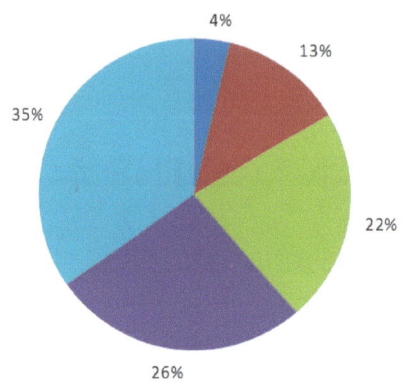

> **Follow-up statement: I sometimes search for information on the Internet using a mobile computer device while driving:**

	Response %	Response Count
Strongly disagree	**59.6%**	**96**
Disagree	14.9%	24
Disagree more than agree	9.3%	15
Agree more than disagree	9.3%	15
Agree	5.0%	8
Strongly agree	1.9%	3

(Source: http://www.surveymonkey.com/MySurvey_Responses.aspx, 05/24/2012)

Mean: 26.83	Median: 15.0	Mode: 15.0

Texting

HOW MUCH EFFECT DOES USING A MOBILE COMMUNICATIONS DEVICE WHILE DRIVING HAVE ON DRIVING ABILITY – TEXTING

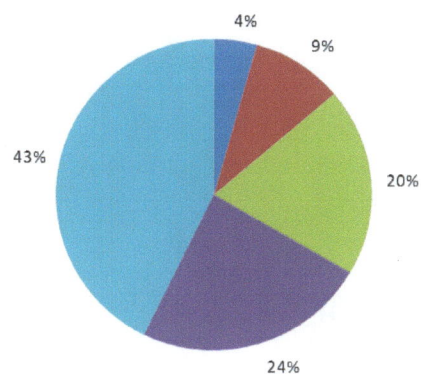

> **Follow-up statement: I sometimes text using a mobile computer device while driving:**

	Response %	Response Count
Strongly disagree	**44.1%**	**71**
Disagree	11.8%	19
Disagree more than agree	12.4%	20
Agree more than disagree	11.2%	18
Agree	18.6%	30
Strongly agree	1.9%	3

(Source: http://www.surveymonkey.com/MySurvey_Responses.aspx, 05/24/2012)

Mean:	26.83	Median:	19.5	Mode:	N/A

E-Mailing

HOW MUCH EFFECT DOES USING A MOBILE COMPUTER DEVICE WHILE DRIVING HAVE ON DRIVER ABILITY – E-MAILING

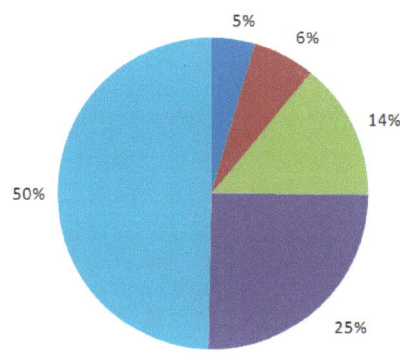

> **Follow-up statement: I sometimes type and send e-mails using a mobile computer device while driving:**

	Response %	Response Count
Strongly disagree	**71.3%**	**114**
Disagree	10.0%	16
Disagree more than agree	7.5%	12
Agree more than disagree	7.5%	12
Agree	3.8%	6
Strongly agree	0.00%	0

(Source: http://www.surveymonkey.com/MySurvey_Responses.aspx, 05/24/2012)

Mean: 26.67	Median:	12.0	Mode:	12.0

Reading

HOW MUCH EFFECT DOES USING A MOBILE COMPUTER DEVICE WHILE DRIVING HAVE ON DRIVING ABILITY - READING

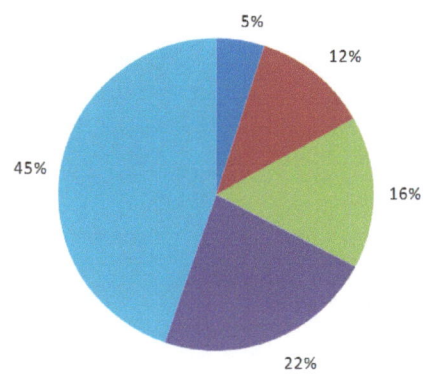

➤ **Follow-up statement: I sometimes read text or e-mails using a mobile computer device while driving:**

	Response %	Response Count
Strongly disagree	**31.9%**	**51**
Disagree	13.1%	21
Disagree more than agree	13.1%	21
Agree more than disagree	17.5%	28
Agree	17.5%	28
Strongly agree	6.9%	11

(Source: http://www.surveymonkey.com/MySurvey_Responses.aspx, 05/24/2012)

Mean: 26.67	Median:	24.5	Mode:	21.0

Downloading

HOW MUCH EFFECT DOES USING A MOBILE COMPUTER DEVICE WHILE DRIVING HAVE ON DRIVING ABILITY - DOWNLOADING

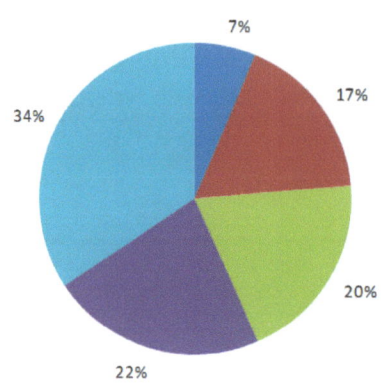

➢ **Follow-up statement: I sometimes download software of data from remote Internet sites using a mobile computer device while driving:**

	Response %	Response Count
Strongly disagree	**71.1%**	**113**
Disagree	18.2%	29
Disagree more than agree	6.9%	11
Agree more than disagree	1.9%	3
Agree	1.3%	2
Strongly agree	0.6%	1

(Source: http://www.surveymonkey.com/MySurvey_Responses.aspx, 05/24/2012)

Mean:	26.5	Median:	7.0	Mode:	N/A

Browsing

HOW MUCH EFFECT DOES USING A MOBILE COMPUTER DEVICE WHILE DRIVING HAVE ON DRIVING ABILITY - BROWSING

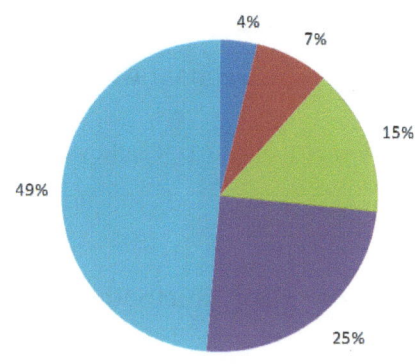

> **Follow-up statement: I sometimes browse web pages using a mobile computer device while driving:**

	Response %	Response Count
Strongly disagree	**70.0%**	**112**
Disagree	12.5%	20
Disagree more than agree	6.9%	11
Agree more than disagree	8.8%	14
Agree	1.3%	2
Strongly agree	0.6%	1

(Source: http://www.surveymonkey.com/MySurvey_Responses.aspx, 05/24/2012)

Mean:	26.67	Median:	12.5	Mode:	N/A

Best Practices in Hands-free/Wireless/Bluetooth Technology

Of immediate importance may be enhancing the array of safety solutions as it pertains to the risks [hazards] of cell phone use. Cimperman (2011) suggested issues of active safety:

1. Consider the specific cell phone absorption rate,

2. Read all the materials that come with the cell phone or training session material (eSafety),

3. Keep cell phone antennas away from pregnant women, babies, and children,

4. Do not keep your wireless devices in your pockets or next to your body when turned on,

5. Use a headset, wireless headphone (like Bluetooth headset/earpiece device), or speaker (eCall) that is endorsed by hands-free laws,

6. Keep cell phone calls short and when you do not need to speak in person, send text messages instead,

7. Use your cell phone only when the signal strength is strong, weak signals increase the output of radio-frequency radiation; and,

8. Avoid using your cell phone inside spaces enclosed by metal, like elevators, subways, trains, planes, and cars (pp. 6-7).

Magnitude of Handheld Bans

A. Handheld Bans for All Drivers (Primary and Secondary Laws)

B. Ban on Cell Phone Use and Texting for All Drivers

C. Ban on Cell Phone Use and Texting for Novice, Bus Drivers, and Government Workers

D. Bans on Texting and Talking for Drivers with Learner's Permit or Intermediate License

E. Ban on Use of Hand-held or Hands-free Phones in School and Construction Zones

F. Preemption Law Prohibits Localities from Enacting Distracted Driving Bans

G. Teens with Probationary Licenses whose Usage Contributes to a Traffic Crash or Ticket

Handheld bans must be followed across jurisdictions around the globe. The magnitude of handheld bans may be addressed through primary and secondary laws as it reflects the above-mentioned drivers: novice, bus, government workers, school, and construction zones for all drivers with a learners permit or immediate license.

Conclusion

The correlation that was uncovered between perception of danger and behavior is important because people need to maintain a level of balance between convenience and safety when using cell phones. The findings of this study reflect the prior work of individuals who have called for more attention on this balance (Constant et al., 2009). When Martin Cooper invented the first mobile cell phone in 1973, he did not initially think about the dangers of motor vehicle drivers who used cell phones as part of a driver related automobile experience. Therefore, the examination of the perception of danger and actual behavior needs to be examined with the big picture in mind for the significance of driver safety solutions on a national and worldwide scale. Since there is a correlation between perceived danger and behavior the wireless industry and public health officials should be building awareness campaigns that seize on this correlation. Therefore, we can influence behavior by influencing perception of danger.

The finding of significance for both gender and age of respondents over the age of 18 hold implications for public awareness campaigns. Constant et al. (2009) mentions that there is not enough education or increased education to spread the awareness of the dangers inherent in motor vehicle cell phone use. Age and gender must be examined to provide the best practices of safety by periodic testing of eyesight and automobile maneuverability. It is important to promote periodic testing to weigh and consider driver distraction prevention efforts and cell phone use. Also, there must be legislation and support from companies that produce the product. Without proper legislation product application may create incidents of fatality. Finally, automobile manufacturers can use this information to develop radio frequency blockers around the drivers in motor vehicles. There may be a need for a physical dialing block on the steering column to address the issue of handheld driver cell phones and driver use. The driver must respect the law or take the risk when it comes to driver safety. The combination of safety solutions should be addressed through attention of age, gender, legislation, and the need for a dialing block device. Practical application of resources may drive innovation of hands-free cell phone devices.

The failure of the other variables to identify any significant correlations between the variables selected for this study has important implications for existing models of education and prevention designed to reduce deaths and personal injury associated with the use of handheld cellular devices. Existing models for public health and safety education hinge on the assumption that behavior is rooted in attitude and driver performance (Constant et al., 2009). The results of this study suggest that this assumption should be revisited regarding these moderating variables and the existence of traffic management solutions. Current campaigns are not tailored to the factors identified in this study; more targeted campaigns may yield better results. The findings of this study align the challenges of Kurt Lewin (1947; Field Theory: Unfreezing, Changing & Refreezing) and Martin Cooper, who found it necessary to make a product and make the product better through research and development. Any time there is the ability to test a product with perceived risk, manufacturers that focus on the healthcare solutions of product use must be considered. Project and time management efforts should be implemented toward aligning safety efforts with everyday application. Since not all products are perfect, users must accept the fact that further research and development will shift the use of impractical applications with progress through updated next generation product solutions.

The telecommunications industry is responsible for providing the public with technologies that enhance life; they are also responsible for encouraging users to avoid endangering the lives of themselves and others through their use. The results of this study affirm some of the existing models of public safety education and the general assumption that behavior is influenced by attitudes. The significance of gender and age as moderating variables suggests there are a variety of different factors that may be significant when designing a cell phone product. Motor vehicle drivers that use innovative products must address the many safety concerns [solutions] needed to maintain safe driver activity that will reduce the number of automobile related deaths. Safety solutions amongst genders may be defined using cell phones and the brain activity that is common in both men and women. Morning, afternoon, and evening cycles of brain activity are different and may provide the best safety solutions through proper grooming, diet, and exercise. As brain activity increases, men and women drivers may be confronted by greater driver distraction and handle this distraction with better cognitive awareness. Managing the inherent driver distractions and interruptions within the vehicle are paramount to driver safety concerns.

The dissertation paves the way for future research on the topic or related topics. As a first formal study for this researcher, this study provides a useful foundation for further investigation into this issue and a basis for guiding professional practice within the wireless communications industry. The goal of this study was to identify ways to help promote the safety and well-being of people who are motor vehicle drivers and users of cellular telephones. These results extend the dialogue focused on identifying strategies that will save a life, prevent an accident, or make drivers aware of potential dangers. Just providing awareness of the behavior or attitudes of motor vehicle related deaths can be a starting point.

The results of this study indicate that there is a correlation between attitude and behavior about driving and the use of handheld devices, and that gender and age are significant moderating variables. The failure to identify a correlation between the other moderating variables suggests that further research is needed into what factors truly will influence safe driving behaviors. These findings provide a basis for improvements in professional practice, as health and safety educators can use these results to inform their awareness campaigns and shape safe driver attitudes among handheld device users with the reflection of users and drivers themselves. The purpose of this study was fulfilled because it contributes insights that exist between the independent variable, perceived dangers of motor vehicle drivers who use cell phones, and the dependent variable, actual cell phone use. As the world of telecommunications continues to evolve and handheld technologies become increasingly spread across regions, the demand for educating and encouraging responsible use of these devices will be paramount as perceived dangers are reduced through the technology of new product development and adaptability to the new products themselves.

The world will be a better place if motor vehicle deaths due to driver related distractions of handheld cell phone use are reduced. Incremental efforts of automobile manufacturers, telecommunications device manufacturers, and traffic management solutions should provide greater product use and educational awareness. Health care professionals must address the issues of innovative technology and the ability for user simplification to the skill set of the driver. Handheld cell phones should be used in moderation and should not be used to distract the motor vehicle driver and his or her passengers, whether distraction is intentional or unintentional.

Therefore, the banning of talking and texting cell phone or smart phone devices need to be regulated nationally and motor vehicle voice activation computers should provide an array of greater driver safety solutions to meet both the passenger and driver associated needs. The value of increasing awareness to change behavior and driver attitudes will fuel product safety campaigns and drive down handheld cellular telephone device use while in the motor vehicle.

Full Dissertation Study Demographics

- 42.7% had income $50 K to $100 K; while 37.2% had under 50 K and 21.3% had over 100 K.
- 67.1 % of males participated in the survey instrument while 32.9% were female respondents.
- 87.2% were Caucasian, while 13.4% were Asian, Black, Hispanic, or Other.
- 93.9% had used a mobile communications device while driving; and 6.1% had not.
- 64.0% were said to be parents while 36.0% were not.
- 35.6% had no children while 15.3% had one, 27.0% had two, and 22.1% had three or more.

- The age group of the participants was 25.0% of 45 to 54; and 20.7% of 55 to 64, 18.3% 25 to 34, 15.2% 35 to 44, 8.5% 18 to 24, 7.9% >64, and 4.3% <18.
- 56.7% were married and living together, while 43.3% were not.
- 93.9% did not have a disability, while 6.1% were disabled.
- 96.3% held a valid driver's license, while 3.7% did not.
- 39.6% held a bachelor's degree, 28% had a high school diploma; 24.4% had a master's degree, 5.5% held a doctorate degree; and 2.4% held no high school diploma.
- 35.6% had no children, while 27% had two, 22.1% had three or more, and 15.3% had one child (Eidelman, 2024, pp. 106-107).

Use of Mobile Phone While Driving

- 59.6% strongly disagree to Internet searches using a mobile communications device while driving.
- 44.1% strongly disagree with texting while driving.
- 71.3% strongly disagree with e-mailing while driving.
- 31.9% strongly disagree with reading while driving.
- 71.1% strongly disagree to downloading while driving, and
- 70.0% strongly disagree to browsing while driving (Eidelman, 2024, pp. 107-108)

Your successful reactions should include:

A. Pull off road to program navigation devices

B. Reduce lane wondering or frequent lane shifts

C. Avoid sharp frequent steering corrections

D.   Moving with the flows of traffic

E.   Reduce sporadic speed changes and sudden breaking

F.   Reduce tailgating or following too closely; and,

G.   Increase distance between you and distracted drivers in high traffic areas

Be aware of traffic laws, signs, and signals. Recognize the need to safely time your stop at a traffic signal. Understand the importance of delayed acceleration from a traffic signal and understand the importance of stopping at railroad crossings and for school buses. Remember, safety involves accident prevention and awareness of the signs around you. Your driving behavior as it relates to speeding and passing should be in your control.

References

Barany, I. & Vu, V. (2007). Central limit theorems for Gaussian polytopes, The Annals of

Probability. *Institute of Mathematical Statistics, 35*(4), 1593–1621.

doi:10.1214/009117906000000791.

Center for Disease Control and Prevention. (2010). *Morbidity and Mortality Weekly Report,*

59(41), 1329-1360.

Cimperman, S. (2011, May/June). Cell phone radiation: Health hazards & how to minimize

your exposure. *Wisdom – New England Edition*, 6-7.

Cohen, J., Ropeik, D., & Myron, K. (2000). Study finds that restricting cell phones while

driving may be premature. *Harvard School of Public Health Press Release*, 1-2.

Retrieved from http://www.hsph.harvard.edu/news/press-releases/archives/2000-

releases/press07242000.html

Constant, A., Salmi, L. R., Lafont, S., Chiron, M., & Lagarde, E. (2009). Road casualties and

changes in risky driver behavior in France between 2001 and 2004 among participants

in the GAZEL cohort. *American Journal of Public Health, 99*(7), 1247-1253. doi:

10.2105/AJPH.2007.126474.

Eidelman, J. A., (2024). *Safety Solutions and Differences in Motor Vehicle Drivers Who Use Cellular Telephones*, KDP Amazon Publishers, Lexington: KY, 1-272.

Lewin, K. (1935). *A dynamic theory of personality – Selected papers.* New York, NY: McGraw-Hill.

Lewin, K. (1936). *Principles of topological psychology.* New York, NY: McGraw-Hill.

Lewin, K. (1947). *Understanding the three stages of change.* New York, NY: Harper & Row. Retrieved from https://www.mindtools.com/pages/article/newPPM

Lewin, K. (2008). *Resolving social conflicts: Field theory in social science.* Washington, DC: American Psychological Association.

Mortazavi, S., Atefi, M., & Kholghi, F. (2011, June). The pattern of mobile phone use and prevalence of self-reported symptoms in elementary and junior high school students in Shiraz, Iran. *Iranian Journal of Medical Sciences, 36*(2), 96-103.

Onwuegbuzie, A. J. & Johnson, R. (2006, Spring). The validity issue in mixed research. *Research in the Schools, 13*(1), 48-63.

Rice, S. & Trafimow, D. (2010). How many people must die over a type II error? *Theoretical Issues in Ergonomics Science, 11*(5), 387-401. doi:10.1080/14639220902853096.

Walsh, S. P., White, K. M., & Young, R. D. (2009). The phone connection: A qualitative exploration of how belongingness and social identification relate to mobile phone use amongst Australian youth. *Journal of Community & Applied Social Psychology, 19*(3), 225-240. doi:10.1002/casp.983

Wang, L., Sato, H., Rau, P., Fujimura, K., Gao, Q., & Asano, Y. (2009). Chinese text spacing on mobile phones for senior citizens. *Educational Gerontology, 35*(1), 77-90. doi:10.1080/03601270802491122.

Profile:

James A. Eidelman, DM/IST, CIE: (617) 694-7243; www.jameseidelman.com

About James: James A. Eidelman is Administrative Assistant at Verizon Communications, a leading provider of wireline, consumer & mass business, wireless, fiber optic and global Internet networks and services. He was named Administrative Assistant on March 28, 2010. From January 4, 2010, until present, Eidelman served as a New England Work-force Planning and Analysis (NE WPA) Clerk, with responsibility for the operations of the company's Force Room activities. He is responsible for NE WPA schedules, overtime, tour swaps and vacation allocation functions.

Eidelman is one of the administrators of today's research management analysis team supporters. He was Receivables Management Collection Center's (RMCC) Service Representative Consultant from 2000 to 2005, until he was moved to the Unlawful Call Center (UCC) as a Service Representative Consultant. He served there from 2005 to 2008. After that he served as Administrative Assistant from January 2010 to present prior to holding a FiOS Sales Specialist position from December 2008 to January 2010.

Before joining Verizon Communications in October 2000, he worked as a Janitor for Temple Emanuel in Newton Center, Massachusetts. He held several part-time and short-term contract positions prior to that working for Aerotek, Inc., AMARR Garage Services, Sara Lee Corporation, Wells Fargo Armored Services, Merchant Services, U.S.A., Exergen Corporation, M.S. Walker, Inc., Computer Ease, Inc., Gerald P. Eidelman, C.P.A., Warehouse Flowers, Inc., the Internal Revenue Service, and the Department of Commerce.

Eidelman is past Security Floor Warden and Union Steward (CWA 1400) while serving for the Unlawful Call Center. Between online telephone calls from customers in the UCC, he facilitated emergency fire drills, from the 18-story building, as well as grievances in support of the UCC employee base. The work Eidelman does now pale in comparison to the work he did at the UCC where he handled calls such as bomb threats, life threats, and harassing calls. Eidelman worked closely with the police, FBI, and the Secret Service along with landline customers within thirteen states and the District of Columbia.

He earned a bachelor's degree in business administration from Boston University (School of Management), and a master's degree in business administration from the University of Phoenix – Flex-Net Program. To further his education, he earned a Doctoral degree (DM/IST) from the University of Phoenix – On-line campus as a Doctor of Management in Organizational Leadership and Information Systems Technology (DM/IST) with a grade point average of 3.84. In 2015, he obtained an Executive Educational Certification from the Massachusetts Institute of Technology (Sloan School of Management). He successfully completed courses in Law, Finance, Management and Leadership.

He has completed ten out of ten modules at the Institute for Chief Information Officer (CIO) Excellence online learning sessions. He passed the Certified Information Executive exam on October 10, 2019. Also, Dr. James has earned a 4.0 GPA (97%, 96%, 98%, 93% 99%, 98%, 98%, 98%, and 98%, respectively) in the Legal Assistant Paralegal, Private Investigator, Medical Billing Specialist, Social Media Strategist, Real Estate Appraiser, Accounting, How to Start Your Own Business, Forensic Science, and Dental Assistant degree programs at Stratford Career Institute; St. Albans, Vermont/Mount-Royal, Canada.

Dr. James A. Eidelman, DM/IST, CIE, Administrative Assistant – January/February 2024

www.jameseidelman.com +1-617-694-7243; JAEidelman@email.phoenix.edu